CONCEPT CARS

Designing the Future

Publications International, Ltd.

Louis Weber, CEO
Publications International, Ltd.
8140 Lehigh Avenue
Morton Grove, IL 60053

ISBN: 978-1-64558-784-2

Manufactured in China.

8 7 6 5 4 3 2 1

Let's get social!

⊙ @Publications_International   🅕 @CollectibleAutomobile

🅕 @PublicationsInternational   🅕 @cgautomotive

www.pilbooks.com

With gratitude to these photographers and car owners:
1992 Jaguar XJ220: photography by Mirco DeCet; **1964 Porsche 904 GTS**: photography by Doug Mitchel; **1953 Cunningham C-3**: photography by Doug Mitchel, owner Thomas R. Coady, Jr.; **1952 Maserati A6GCS**: photography by Doug Mitchel, owner The Brumos Collection; **1938 Talbot-Lago**, photography by Nicky Wright, owner Brooks Stevens Museum

Special thanks to following manufacturers who supplied imagery.
Aston Martin Lagonda Limited, BMW Group, Bugatti Automobiles, S.A.S., Daimler AG, Ferrari, S.P.A., Fiat Chrysler Automobiles, Ford Motor Company, General Motors Company, Hyundai Motor Company, Jaguar Land Rover Limited, Koenigsegg Automotive AB, Automobili Lamborghini S.p.A., Mazda Motor Corporation, McLaren Automotive, Nissan Motor Corporation, Porsche AG, Rolls-Royce Motor Cars Limited, Shelby American, Tesla Inc., Toyota Motor Corporation, Volkswagen AG

Additional images from Shutterstock.com

TABLE OF CONTENTS

INTRODUCTION

Concept Cars explores the inventive world of show vehicles—the forward-looking auto-show display cars that have long been used by automakers to excite the public and test new ideas. In addition to concept cars, you'll also discover vehicles that illustrate what happens when early designs are greenlit into production runs. Take the case of the 2020 Corvette Stingray whose mid-engine layout realized a dream of the late godfather of the Corvette, Zora Arkus-Duntov. The genesis of the 2020 model dates all the way back to the 1973 Aerovette concept car that nearly went into production in 1980.

While many show cars hew toward the non-traditional, other concept cars have predicted technologies that are now commonplace among today's passenger vehicles, such as rearview cameras and automatic emergency braking. Dream cars (as they're also known) not only foretold the future, but prepared the public for coming design trends and features. Although some concept vehicles were pure flights of fancy, others tested styling themes that would appear on production cars a few years later.

Within the pages of *Concept Cars* you'll "meet" cool vehicles like the 2020 Aston Martin Valkyrie hypercar, 2017 Lamborghini Terzo Millennio electric supercar, the self-referential 2011 BMW 328 Hommage that paid tribute to BMW's early years, and many more.

Just as recent concept vehicles helped point to the promise of electric power like never before, today's dream cars continue to reinvent luxury, performance, and head-turning design. Modern technology is now poised to make self-driving cars and other optimistic prophecies of the twentieth century feasible, and concept cars are still tantalizing us with what tomorrow could bring.

The 2021 Roma's design harkens back to the marquee's powerful and speedy coupes of the past, but the Roma provides a spacious interior that Ferrari coupes have never known before.

The Roma is a throwback to graceful, grand touring Ferraris of the Fifties and Sixties. It doesn't have the 986-horsepower rating of the SF90 and it can't match the lap times of the mid-engine Ferraris.

Like many modern sports cars, the Roma does not offer a manual transmission. The 8-speed dual-clutch automatic transmission allows for faster shifting between gears and higher engine efficiency.

The Roma is based on the front-engine Ferrari Portofino convertible and shares a 612-hp twin-turbocharged 3.9-liter V-8 with the convertible. The 8-speed dual-clutch automatic is mounted in the rear to help achieve a balanced 50/50 weight distribution. Ferrari claims the Roma can accelerate 0-62 mph in 3.4 seconds and reach a top speed of 199 mph.

The Roma's interior contains your standard gauges on the digital instrument panel, including climate control and a passenger-side control screen as well.

The proportions of the Roma suggest classic Ferraris such as the 275 GTB or 365 GTB/4 Daytona. Still, the slender headlights and perforated grille are pure twenty-first century. Even at a stiff $222,620, Roma is considered an entry-level Ferrari.

2020 CHEVROLET
CORVETTE STINGRAY

The late Zora Arkus-Duntov was the godfather of the Chevrolet Corvette during its first three decades and pushed for a mid-engined 'Vette in the Seventies. The Aerovette prototype came close to reaching production for 1980, but Chevrolet decided to continue with the conventional front-engine Corvette instead. Duntov's dream of a mid-engined Corvette was finally fulfilled for 2020.

The mid-engine layout had long been the norm for exotic supercars. With the 2020 Corvette, Chevrolet joined the club. In spite of its exotic-car specs, the new Corvette Stingray started at a reasonable $59,995.

The 2020 Corvette featured a new automatic dual-clutch transmission that removed the clutch pedal and gear shift from the interior design. Many Corvette enthusiasts were dismayed by the removal of the manual transmission, but the Corvette's design and breakneck power surely made up for the supposed shortfall.

The Corvette's control center may seem intimidating, but it featured your standard array of digital gauges for monitoring the car's performance. The defining ridge of the center console featured the car's climate control options.

2020 ASTON MARTIN
VALKYRIE

Aston Martin partnered with Red Bull Racing of Formula 1 fame to develop an all-new hypercar called the Valkyrie. When first revealed in July 2016, the car was referred to by its codename, AM-RB 001 ("AM" for Aston Martin, and "RB" for Red Bull). In March 2017, the Valkyrie name—inspired by the Norse mythological figure—was announced.

The Valkyrie was designed by Red Bull Racing's Chief Technical Officer, Adrian Newey. The form-follows-function exterior styling is a result of Newey's focus on downforce and aero-dynamic efficiency.

Noteworthy styling details included exposed headlight assemblies and mounting brackets, a chemically etched aluminum Aston Martin badge on the car's nose that was thinner than a human hair and buried under a coat of clear lacquer, and what was claimed to be the world's smallest center-mount brake light.

Aston Martin claimed the car "boasts truly radical aerodynamics for unprecedented levels of downforce in a road-legal car," and said that much of the car's aerodynamic downforce was generated by the design of the underside Venturi tunnels.

At the Geneva Motor Show in March 2018, Aston Martin and Red Bull introduced the track-only Valkyrie AMR Pro. Weighing in at 1,000 kilograms (about 2,200 pounds), the car featured numerous weight-saving measures, including lighter-weight carbon-fiber bodywork, polycarbonate windows, carbon-fiber suspension wishbones, race seats, and a lighter exhaust system. In addition, interior comfort and convenience pieces were deleted.

2018 MCLAREN
720S

Surrey, England-based McLaren Automotive launched in 2010, and its first road car was the 2011 McLaren 12C. At the Geneva Motor Show in March 2017, McLaren Automotive introduced its second-generation Super Series car, the 2018 McLaren 720S. Like all other McLaren road cars starting with the F1, the 720S was built on a carbon-fiber tub; this iteration was called the McLaren Monocage II. Bodywork was a combination of carbon fiber and aluminum. The 720S rode a 105-inch wheelbase, was 179 inches long, and had a curb weight of 3128 pounds.

27

Exterior styling was an evolution of the McLaren look. Highlights included slim roof pillars that enhanced occupant visibility and taut "shrink-wrapped" bodywork.

The dihedral doors were double skinned, which allowed them to hide inner ducting that channeled air to the car's radiators and eliminated the need for open side intakes.

The mid-engine, rear-drive 720S ran McLaren's M480T twin-turbo-charged 4.0-liter V-8, which was rated at 710 horsepower and 568 pound-feet of torque. The engine was mated to a 7-speed dual-clutch SSG (seamless-shift gearbox) transmission.

2018 LEXUS
LC 500

Lexus called its LC 500 coupe the luxury brand's flagship. It also said it's a concept car come to life, and in this case it's the LF-LC concept first shown at Detroit's North American International Auto Show in early 2012.

The first thing most people noticed about the LC 500 was its expressive styling. The classic long-hood/short-deck coupe proportions were present, and the racy roofline tapered rearward.

The LC 500's extra-long hood hid a 5.0-liter naturally-aspirated 32-valve V-8 engine. It was rated at 471 bhp and 398 pound-feet of torque. Horsepower peaked at a lofty 7100 rpm, with a redline set at 7300 rpm.

The engine mated to a 10-speed automatic transmission and was fitted with a standard active exhaust system that allowed the driver to select the level of exhaust sound. Its rambunctious exhaust added a surprising muscle car vibe. Lexus claimed the LC 500 would accelerate from 0 to 60 mph in 4.4 seconds.

As a 2+2 coupe, the LC 500 was dramatically styled and luxuriously appointed inside. However, that style came at the price of nearly nonexistent rear-seat legroom.

Muscular fenders dramatically bulged outward, and bodyside scoops provided an additional sporting touch. Chrome moldings framed the sides of the standard glass or optional carbon-fiber roof panel. Other highlights included elaborately-detailed headlight and taillight assemblies and dramatic pop-out door handles.

The rear-drive LC ran on a 113-inch wheelbase, was 187.4 inches long, and stretched 75.6 inches wide. Curb weight was a substantial 4280 pounds.

At introduction, the LC 500 was priced from $92,000. Individual extras included a cold weather pack with a heated steering wheel and windshield deicer, color heads-up display, and a limited-slip rear differential.

Option groups were a Touring Package, Sport Package, Sport Package with Carbon Fiber Roof, Performance Package, and Mark Levinson-brand audio.

The Lexus LC 500 could easily cost north of $100,000, but it packed exotic looks, beautifully finished interior, and, of course, ample V-8 power.

2018 LAGONDA
VISION CONCEPT

Aston Martin bought the Lagonda make in 1947 and often used the Lagonda name on its sedan models.

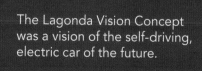

The Lagonda Vision Concept was a vision of the self-driving, electric car of the future.

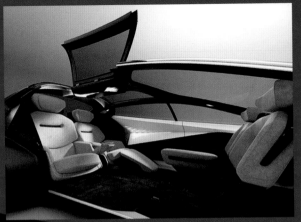

With the battery mounted under the floor and no need for a bulky gas engine, the Vision Concept devoted most of its space to the passenger compartment.

2018 JEEP
4SPEED

Jeep typically introduced modified versions of its products at its Easter Jeep Safari in Moab, Utah.

For 2018, Jeep showed a special 4SPEED that was 22 inches shorter and 950 pounds lighter than the production Wrangler.

The changes made the Jeep even more agile off road and, of course, faster.

2018 HYUNDAI
LE FIL ROUGE

The 2018 Le Fil Rouge concept car introduced Hyundai's "Sensuous Sportiness" design theme that was expected to influence future Hyundai models.

Sensuous Sportiness was defined as the harmony between four fundamental elements in vehicle design: proportion, architecture, styling, and technology.

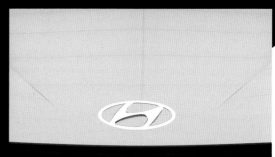

The Le Fil Rouge's interior used revitalized wood and high-tech fabrics.

56

2017 TESLA
ROADSTER

Before Tesla built sleek electric sedans, the carmaker launched with a small electric sports car.

The first Tesla Roadster went out of production in 2011 and the concept car displayed in 2017 pointed the way for a second-generation Roadster.

Tesla claimed the new Roadster would
accelerate 0 to 60 mph in 1.9 seconds
and have a top speed of 250 mph.

2017 NISSAN
VMOTION 2.0

The 2017 Vmotion 2.0 concept car suggested the direction of Nissan design for future sedans.

The Vmotion 2.0 also showcased the latest version of Nissan's ProPilot automated driving system that could drive the car within a selected lane.

When ProPilot was operating, badges on the front and rear of the car glowed.

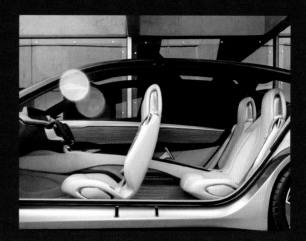

2017 MAZDA
VISION COUPE

Automakers often call a sleek four-door sedan a "coupe" because of the car's smooth, coupe-like roofline. A case in point is the 2017 Mazda Vision Coupe.

The Vision Coupe is a continuation of the design theme of the red 2015 Mazda RX-Vision concept—a car that was a true coupe.

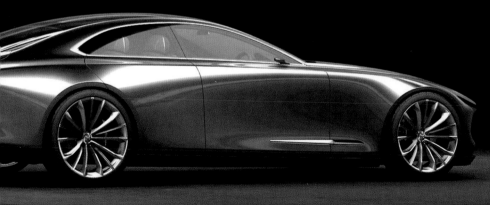

The concept's exterior features a "one motion" form exuding a sense of speed.

2017 BUGATTI
CHIRON

There are supercars, and then there are supercars, and the Bugatti Chiron definitely qualifies as the latter. The Chiron's quad-turbocharged 8.0-liter W-16 engine developed an astonishing 1500 horsepower and 1180 pound-feet of torque—enough for a 0-60 mph time of 2.4 seconds and an electronically limited top speed of 261 mph.

A Haldex all-wheel-drive system got all that power to the road with the least amount of drama. The Chiron was more refined and quiet than its predecessor, the Bugatti Veyron, and was also surprisingly docile at low speeds.

The Chiron set a world record by accelerating 0 to 400 km/h (248 mph) and then braking to a stop in 41.96 seconds. That record was quickly bettered by Koenigsegg, but it was still a remarkable accomplishment.

Base price for the Chiron was around $3,000,000. As expected at that price, there was no visible plastic in the cockpit—only top-grade leather and metal.

The Bugatti Chiron was presented at the Geneva Motor Show in March 2016.

2017 FERRARI
LA FERRARI

Ferrari would probably be the last make expected to sell a hybrid, but the LaFerrari was no ordinary hybrid—it combined a 788-horsepower V-12 with a 161-hp electric motor for a total 949 hp.

Introduced for 2013, the LaFerrari hybrid was even faster than Ferrari's celebrated 2003-04 Enzo model—it could rocket from 0-60 mph in less than three seconds and hit a top speed of 217 mph.

The LaFerrari's price tag was a cool $1.4 million, but Ferrari had no trouble selling 500 coupes, as well as 210 Aperta convertible versions. The final LaFerrari Aperta was sold at an auction in 2017 for $9.96 million, with proceeds supporting the Save the Children charity.

The weight of the low-mounted battery packs lowered the car's center of gravity and improved handling.

2017 KOENIGSEGG
AGERA

Koenigsegg's Agera model debuted in 2010. The company says "Agera means 'to take action' in Swedish." Of interest here is the Agera RS, which was originally shown at Geneva in March 2015. They claimed it to be "designed as a road-legal car with a track focus."

The company announced that twenty-five cars would be built, with ten of them pre-sold prior to the first showing. Prices likely started well north of a million U.S. dollars.

The Agera's tub was hand assembled in Koenigsegg's workshop from carbon fiber and an aluminum honeycomb core. Bodywork is carbon fiber as well. The car is approximately 169 inches long, runs a 104.8-inch wheelbase, and is 44.1 inches tall. Curb weight is 3,075 pounds. The engine was a Koenigsegg-developed 32-valve twin-turbocharged 5.0-liter V-8. It was rated at 1,160 horsepower and 944 lb-ft of torque.

An optional "1MW" version of the engine raised the stakes to 1,360 horsepower and 1011 lb-ft of twist.

On October 1, 2017, a U.S.-spec Agera RS with the 1MW engine driven by a factory driver, accelerated from 0 to 400 kilometers an hour (approximately 248.5 mph) and then decelerated back to 0 in 36.44 seconds.

Geneva, 2015. One of the Agera's defining features was the firm's Dihedral Sychrohelix door hinge. The company described this hinge as opening the "door outwards and upwards in one smooth, sweeping motion."

Factory supplied performance numbers with the 1MW engine were 0–62 mph in 2.8 seconds and 0–186 mph in 12.3 seconds.

2017 LAMBORGHINI
TERZO MILLENNIO

The 2017 Terzo Millennio is Lamborghini's vision for the supercar of the future.

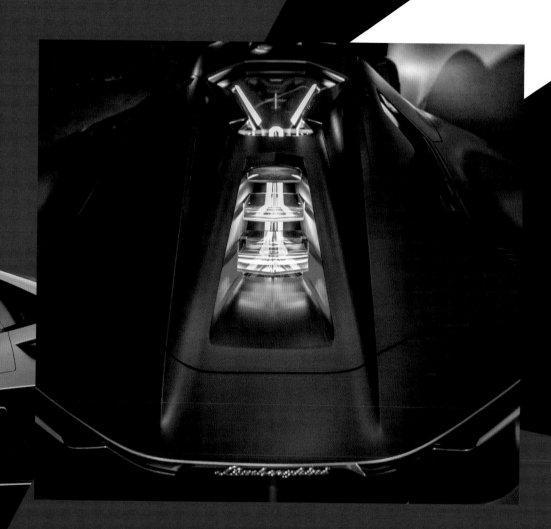

Lamborghini teamed up with the Massachusetts Institute of Technology to develop a concept for an electric sports car that used a supercapacitor instead of a battery to store energy.

An electric motor was incorporated in each wheel to eliminate the need for heavy drive shafts.

Modern Lamborghini design hallmarks are present, like the Y-shaped design of the headlights and taillights.

2017 FORD
GT

To celebrate the 50th anniversary of the Ford GT40's historic 1966 Le Mans victory, Ford designed an all-new GT racecar for the 2016 24 Hours of Le Mans. The new GTs did their legendary forebears proud, finishing first, third, fourth, and ninth in the GTE Pro Class.

The production GT was closely related to the racing version, and lacked the luxury features often found on other high-end supercars. The GT's cockpit was a tight fit for two passengers, and cargo room was almost nonexistent. The payoff was reduced weight, with racecar-like performance and handling.

The EcoBoost 3.5-liter V6 shared its basic engine block with the Ford F-150, but developed 647 horsepower and was capable of traveling from 0-60 mph in 2.9 seconds and reaching a top speed of 216 mph, according to *Car and Driver*.

The price was around $450,000, and Ford planned to build 1000 examples over the 2017–2020 model years.

2016 VOLKSWAGEN
BUDD-E

For many years Volkswagen introduced concepts that suggested that it would build a retro vehicle inspired by its iconic Microbus. In 2016, the BUDD-e concept was an electric version of the Microbus theme.

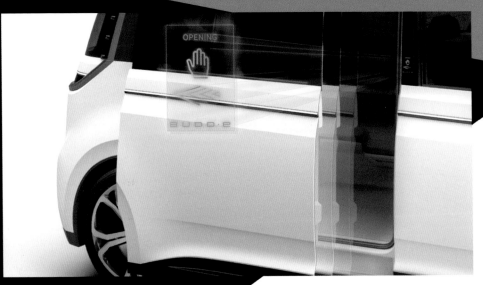

The concept vehicle's interior was innovative and notably forward-looking. The BUDD-e featured a new type of operation system, information via voice, touch and gesture control, and large displays.

The BUDD-e was ultimately passed over and an electric vehicle with more Microbus-like styling was expected to go into production.

2016 ROLLS-ROYCE
103EX

The 2016 Rolls-Royce 103EX was the luxury automaker's concept of a driverless car for the year 2040.

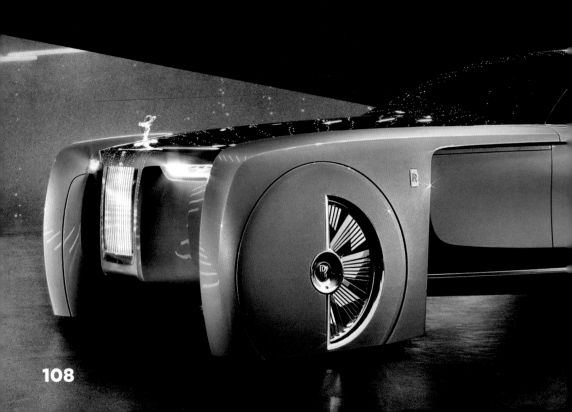

An artificial-intelligence assistant bore the name Eleanor, in reference to Eleanor Thornton, the woman said to have been the model for the Rolls-Royce Spirit of Ecstasy mascot.

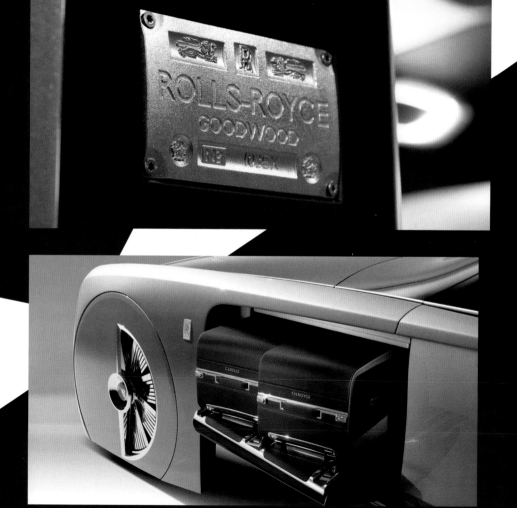

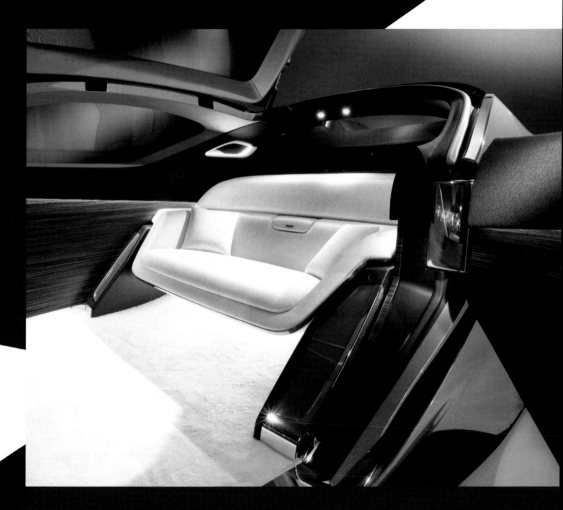

The vast interior contained a silk couch for only two passengers, and hand-twisted silk carpet for your feet to rest on.

2016 VISION
MERCEDES-MAYBACH 6

The 2016 Vision Mercedes-Maybach 6 was a long, sleek coupe concept from Mercedes-Benz's upper-crust niche brand.

Although 18.5 feet long, the electric-powered car was said to be capable of 0 to 62 mph in under four seconds and had a range of more than 200 miles.

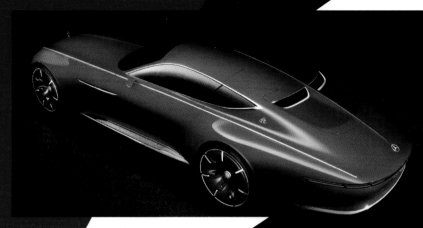

117

The 2+2-seater was a homage to aero coupes and its extended, round "boat tail" even recalled a luxury yacht.

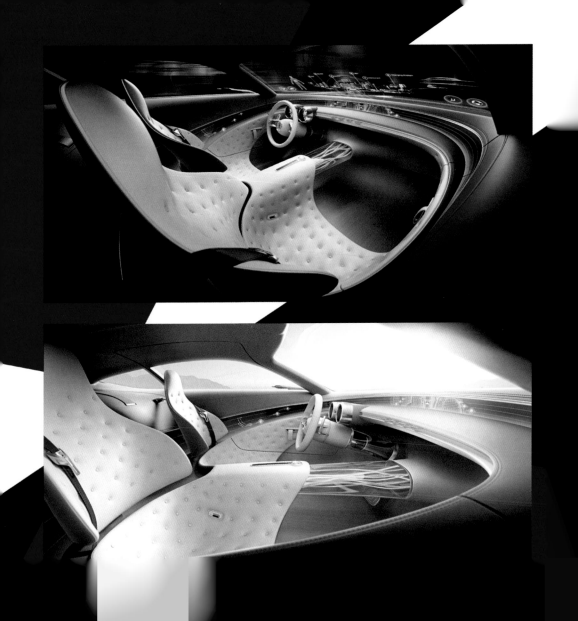

2016 BUICK

AVISTA

Coupes were rare on the automotive landscape in 2016 and hardtop coupes (with pillarless side windows) were even less common. Enter the Avista.

The Buick Avista concept hardtop coupe concept shared a platform with the Cadillac ATS and Chevrolet Camaro and seemed feasible as a production car.

The fading pattern on the seats, console and doors were inspired by waves receding at a beach's edge.

Although well received, the Avista didn't make it to the showroom floor.

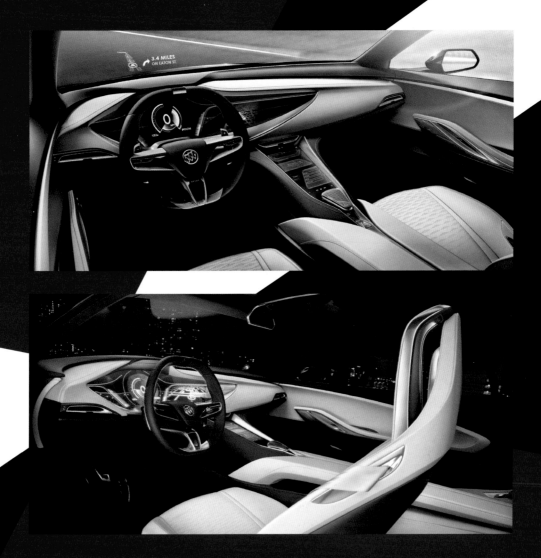

2016 MERCEDES-BENZ
F 015 LUXURY IN MOTION

The 2015 Mercedes-Benz F 015 Luxury in Motion concept vehicle was Mercedes' prediction of transportation in the year 2030 when autonomous driving would be commonplace.

The Luxury in Motion's front seats could swivel to the rear to aid conversation with the rear passengers.

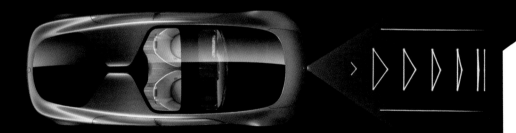

Six screens provided information about the vehicle and a connection with the outside world.

2014-2019 BMW

I8

Beginning with the 2009 introduction of its Vision Efficient Dynamics concept vehicle, BMW began teasing the idea of an exotic sports car with a high-tech, eco-conscious focus. After exploring the concept further (and introducing the i8 name) on a couple subsequent concept vehicles, the company officially committed to a production model.

The i8 was 184.6 inches long, 76.5 inches wide, and 51 inches tall, with a curb weight of 3285 pounds.

The production i8 launched as a 2014 model, with unorthodox styling that covered an equally unconventional plug-in-hybrid powertrain. And almost all of the concept vehicles' outlandish features—most notably the scissor-wing doors—made the jump from the show floor to the showroom.

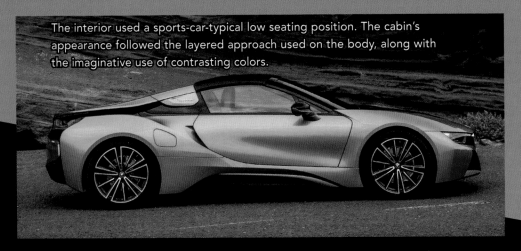

The interior used a sports-car-typical low seating position. The cabin's appearance followed the layered approach used on the body, along with the imaginative use of contrasting colors.

The driver could choose from five driving modes using the Driving Experience Control switch and eDrive button.

Coupe prices started at $147,500, and the new Roadster model had a base price of $163,300.

The starting price was steep ($135,700 in the United States), but not unreasonable compared to similar luxury exotics.

BMW quoted the i8's 0-60-mph time at 4.2 seconds, and the EPA's gas-electric mileage estimate was 76 MPGe in combined city/highway driving.

2013 CADILLAC
ELMIRAJ

Cadillac debuted its Elmiraj concept car at the Pebble Beach Concours d'Elegance in 2013.

The large coupe didn't enter production, but displayed some future Cadillac style cues.

Under the hood was a twin-turbocharged 4.5-liter V-8 with an estimated 500 horsepower.

2011 ASTON MARTIN
CC100 SPEEDSTER

Aston Martin celebrated its 100th anniversary in 2013 with a
CC100 Speedster concept car that was inspired by the 1959
Le Mans winning Aston Martin DBR1 racecar.

The CC100 packed a 6.0-liter V-12 paired with a
six-speed automated-manual transmission. The
interior combined carbon fiber and leather.

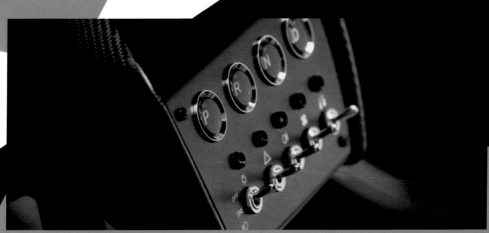

2011 CADILLAC
CIEL

The 2011 Cadillac Ciel was a large, ultra-luxury, four-door convertible riding on a long 125-inch wheelbase.

The Ciel had a hybrid powertrain with a 425-horsepower twin-turbocharged V-6 paired with an electric motor.

Cadillac's Ciel concept car was unveiled at the Pebble Beach Concours d'Elegance.

2011 BMW
328 HOMMAGE CONCEPT

In 2011, BMW unveiled the BMW 328 Hommage concept car as a tribute to the marque's 328 sports car that debuted 75 years earlier.

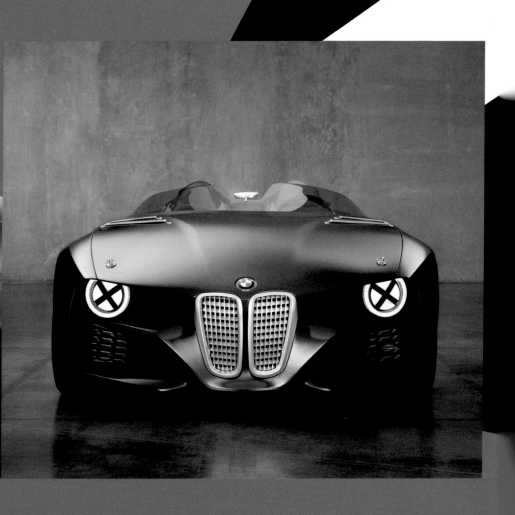

The original 328 helped establish BMW as a maker of sporting machines and the 328 Hommage harks back to the original with a double kidney-shaped grille and leather hood straps.

2010 JAGUAR
C-X75

The 2011 Jaguar C-X75 supercar concept was a radical hybrid that combined electric motors with a turbine engine.

163

Jaguar announced plans to build 250 C-X75s, but the electric motors would have been combined with a conventional internal-combustion motor instead of the turbine.

Projected performance figures were: 0-60 mph in less than three seconds, a top speed in excess of 200 mph, and a 37-mile electric range.

Jaguar canceled the C-X75 after building five prototypes.

2009-2012 CADILLAC
CTS-V

Cadillac's second-generation CTS-V debuted for '09 packing the most-powerful production engine that Cadillac had ever offered: a revised Chevrolet Corvette ZR1 supercharged and intercooled LS9 6.2-liter V-8.

CTS-V offered a choice of transmissions: a Tremec TR-6060 six-speed manual or General Motors' Hydramatic 6L90-E six-speed automatic with steering-wheel-mounted buttons for manual shifting. A handful of performance-themed design alterations differentiated the V from its regular-line CTS siblings.

169

Cadillac claimed the CTS-V could accelerate from 0-60 mph in 3.9 seconds and do the quarter-mile in 12 seconds at 118 mph, but magazine road testers generally couldn't replicate those numbers. Outside of a noticeably stiffer ride and a menacing exhaust note, the CTS-V gave up little to its tamer siblings in mundane stop-and-go driving.

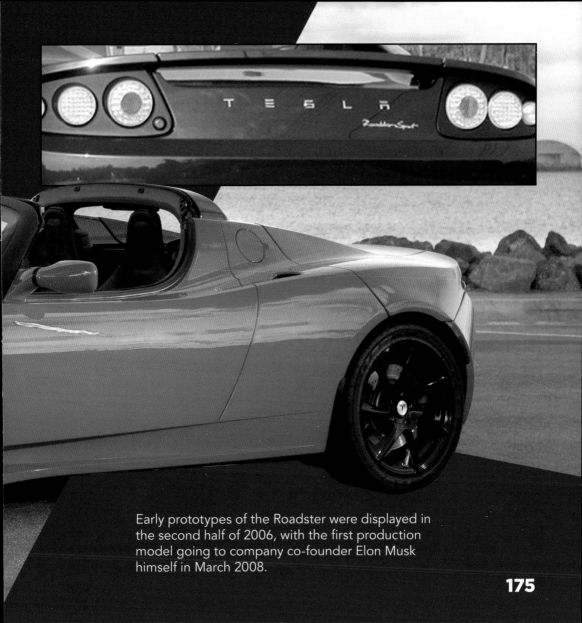

Early prototypes of the Roadster were displayed in the second half of 2006, with the first production model going to company co-founder Elon Musk himself in March 2008.

The Roadster's sleek carbon-fiber body was made in France, then shipped to Lotus of England for placement on a specially designed chassis. After assembly, the body/chassis was shipped to Tesla's shop in Menlo Park, California, to have the battery pack (6,831 lithium-ion laptop-computer batteries), powertrain, and electronic controller installed.

The first Roadsters had a 248-horse-power motor. By 2010, horsepower had been increased to 288.

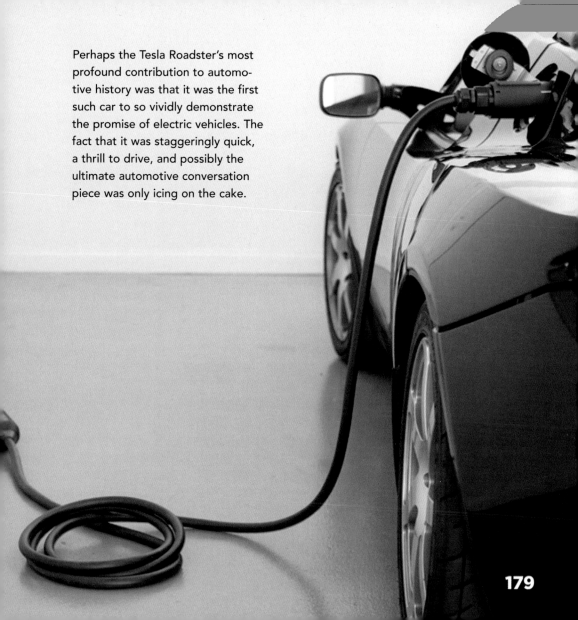

Perhaps the Tesla Roadster's most profound contribution to automotive history was that it was the first such car to so vividly demonstrate the promise of electric vehicles. The fact that it was staggeringly quick, a thrill to drive, and possibly the ultimate automotive conversation piece was only icing on the cake.

2007 CHEVROLET
VOLT

General Motors eased into the electric-car market
with the Chevrolet Volt, a car with an electric motor
and a "range extender" gasoline engine to generate
electricity when the battery ran out of charge.

The Volt's twenty-one-inch wheels and exterior design recall classic Chevrolet performance and two vehicles in particular, the Camaro and the Corvette.

The 2007 Chevrolet Volt concept car was more flamboyant than the more practical production version.

2006 BMW
MILLE MIGLIA

In 1940, a BMW 328 with an advanced aerodynamic body won the Italian Mille Miglia race. BMW paid tribute to that important victory with the 2006 BMW Concept Coupe Mille Miglia.

The Mille Miglia's styling put a modern spin on the prewar racecar. Under the lightweight body of carbon fiber and aluminum was the drivetrain of a BMW Z4 Coupe.

Without conventional doors, an airplane-style canopy hinges at the rear to expose a snug two-seat interior and intricate roll cage.

Design highlights included round headlamp graphics, a split windshield, and a rear end featuring an extended overhang and gently sloping tail.

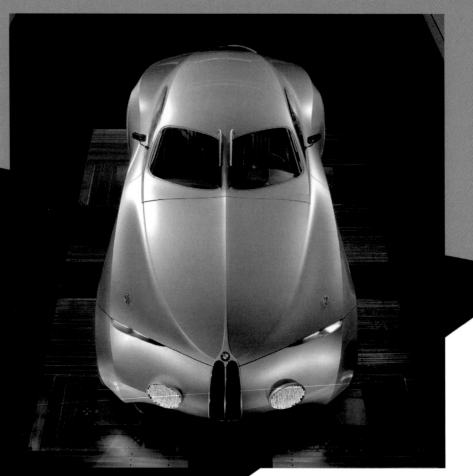

2005 BUGATTI
VEYRON

Bugatti's twenty-first century renaissance began with the Veyron and its promised 1001-horsepower V-16. Federalized models actually delivered 987 hp, roughly double Viper's power output.

Under Volkswagen control, the latest revival of the storied Bugatti name was free of the financial limitations that killed a 1990s rebirth effort.

Production began for 2005, and was limited to 50 vehicles annually. Veyron production ended in 2015.

2005 FORD SHELBY
GR-1

The 2005 Ford Shelby GR-1 was to be the
successor to the 2005-06 Ford GT supercar.

From behind, the GR-1 sports a Kamm tail paired with quad exhaust ports, additional vents, and specialized vertical tail lamps.

The GR-1 concept car was powered by a 605-horsepower 6.4-liter V-10, but a production version would have used a lighter V-8 with similar horsepower.

Ford decided not to produce the GR-1 and supercar enthusiasts had to wait until 2017 for a new Ford GT.

2004-2005 CADILLAC
XLR

The Cadillac XLR that debuted in 2004 was a suave Mercedes SL fighter based on Chevrolet Corvette bones. Yet the XLR looked nothing like a Corvette or the SL, had a pure Cadillac heart, and was plenty fast. The XLR was basically the for-sale version of the 1999 Evoq concept car, the first public hint that Cadillac's future in the twenty-first century would not be like Cadillac's past.

The XLR was not a Corvette in a Cadillac suit, though the similarities were undeniable. For starters, both had front-mounted V-8s, rear transaxles, and composite-paneled bodies. Apart from looks and intended audience, it was the powertrain that most separated the GM twosome.

The XLR differed from the Corvette in many ways, but a major distinction was the XLR's mandatory five-speed automatic transmission with manual shift gate vs. Corvette's four-speed automatic or six-speed manual.

The XLR also replaced the 'Vette's conventional convertible top with a disappearing hardtop and added luxury-class equipment.

The critics praised most everything about it, from the smooth, quiet acceleration—0-60 mph took no more than the factory-claimed 5.9 seconds—to satisfying backroad agility, controlled but comfortable ride, and terrific build quality.

In all, the XLR was arguably the most impressive new Cadillac since the '67 Eldorado.

2003 FERRARI

ENZO

The crown jewel of the Ferrari lineup, the limited-production Enzo came and went in the blink of an eye. Applying the company's famous "demand minus one" formula, Ferrari built just 399 of the stunning coupes—all in 2003 and 2004. Extracting 660 horsepower from a 6.0-liter midship-mounted V-12, and weighing less than 3000 pounds, performance was eye-popping.

According to *Road & Track* magazine, the Enzo sprang from 0-60 mph in a scant 3.3 seconds, and topped out at nearly 220 mph. Equally breathtaking was the Enzo's price, about $650,000. The only available transmission was a 6-speed clutchless "sequential" manual unit with steering-wheel-mounted paddle shifters.

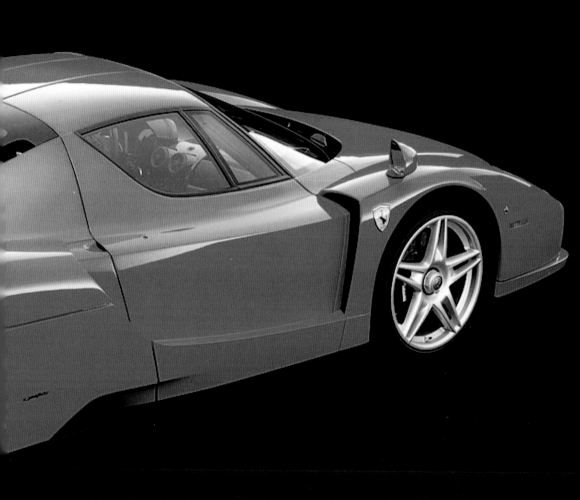

Extensive use of exotic materials helped make Ferrari's lithe dancer the welterweight it was. Body panels were formed of a carbon fiber and Nomex "sandwich," while the chassis and tub were formed of carbon fiber.

Guaranteeing the Enzo's rarity stateside, Ferrari shipped only 100 cars to America.

2003
Ferrari
Enzo

2003 CADILLAC
SIXTEEN

The 2003 Cadillac Sixteen recalled
the marque's flagship of the Thirties,
but the styling and engineering
looked to the future.

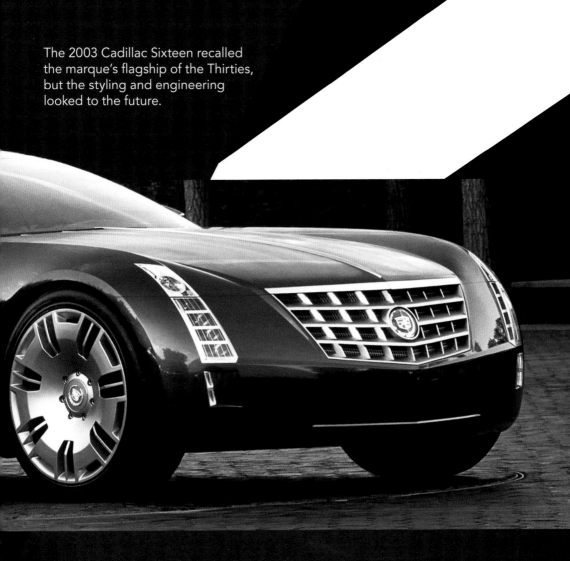

The Cadillac Sixteen's aluminum hood is quite long, giving it marvelous dash-to-axle dimension. Its wheel arches were designed to accommodate the stunning 24-inch polished aluminum wheels.

The Sixteen was the hit of the auto shows with its dramatic styling and astounding specifications. However, Cadillac decided not to put the Sixteen into production.

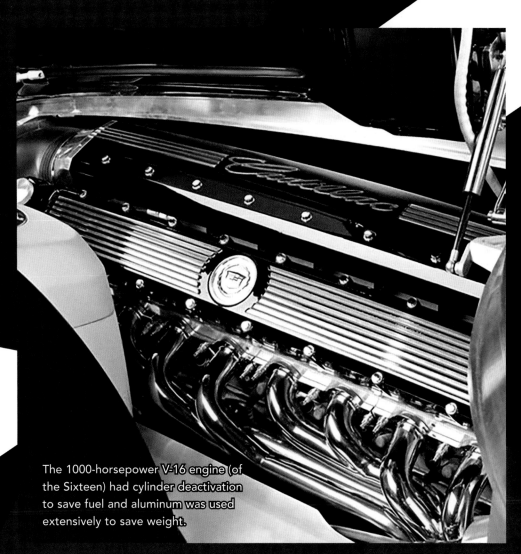

The 1000-horsepower V-16 engine (of
the Sixteen) had cylinder deactivation
to save fuel and aluminum was used
extensively to save weight.

2004 PORSCHE
CARRERA GT

Bearing a name synonymous with sports cars, Porsche was not about to be left out of the supercar renaissance. The stunning Carrera GT arrived on the scene for 2004, and moved promptly to the head of the Porsche class. The GT boasted Porsche's largest ever street-going engine, a 5.7-liter V-10. With 605 bhp on tap, the midship-mounted-engine moved the 3000-pound GT to 60 mph in a factory-claimed 3.9 seconds.

Top speed was reported to be in excess of 200 mph.

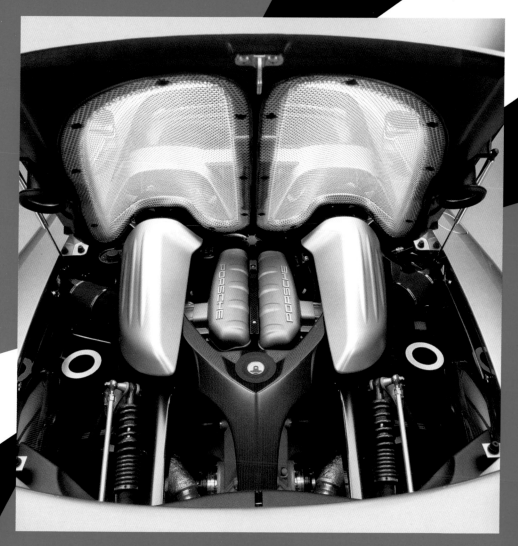

Price: just under $400,000. To make the most of the car's limited storage space, Porsche included a matching five-piece luggage set with each car.

2001 FORD
49

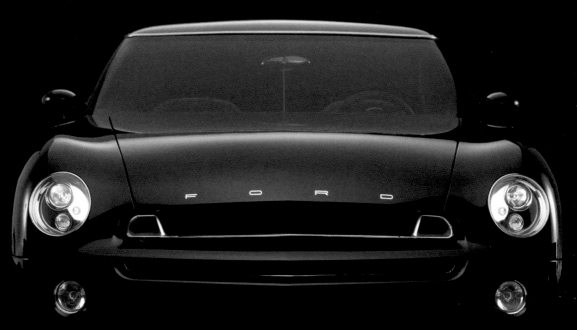

The 1949 Ford's styling and engineering were a huge step forward and a turning point for the company's fortunes. The 2001 Ford 49 concept harked back to the 1949 model in styling, but with a modern flair.

Two words that come to mind are simple and smooth.
The all-glass upper body structure includes totally
concealed pillars and windshield wipers.

The Forty-Nine's interior trended retro, with clean lines and minimal shiny trim.

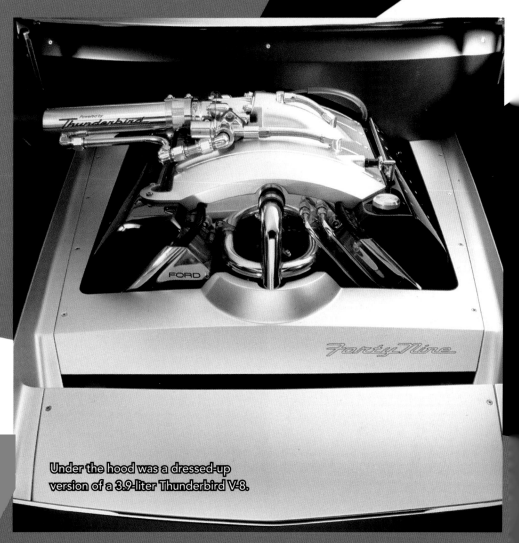

Under the hood was a dressed-up
version of a 3.9-liter Thunderbird V-8.

2001 JAGUAR
XKR SILVERSTONE

In 2001, Jaguar brought out a Silverstone edition of its grand touring XKR coupe and convertible. To XKR's equipment the Silverstone added: Platinum Silver paint, "Silverstone" writ on hood emblem and chrome door-sill plates, bird's-eye maple interior planking instead of the usual burled walnut, charcoal leather upholstery with red stitching, larger-diameter all-disc brakes by Italy's Brembo, and 20-inch wheels wearing bigger tires.

The coupe chassis was further upgraded with a Performance Handling Pack comprising a slightly larger front antiroll bar, a slightly smaller rear bar, higher-rate springs, and steering with a recalibrated electronic control unit and a rack mounted on firmer bushings.

The 2000 XKR's V-8 packed 370 bhp and 387 pound-feet of torque.

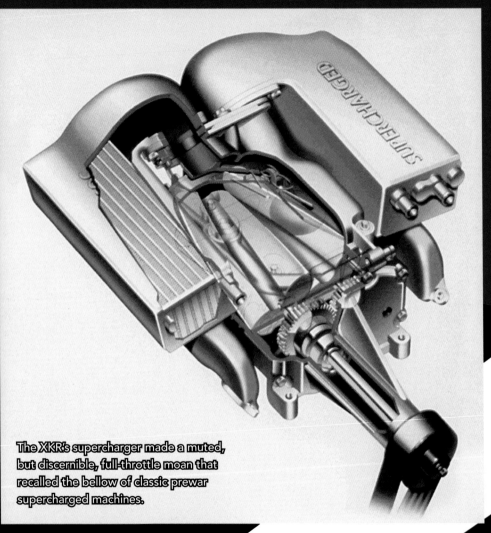

The XKR's supercharger made a muted, but discernible, full-throttle moan that recalled the bellow of classic prewar supercharged machines.

The standing quarter-mile came up in as little as 13.7 seconds at 105 mph. Top speed was electronically limited to 156 mph, but disconnecting the governor might have increased that to 170.

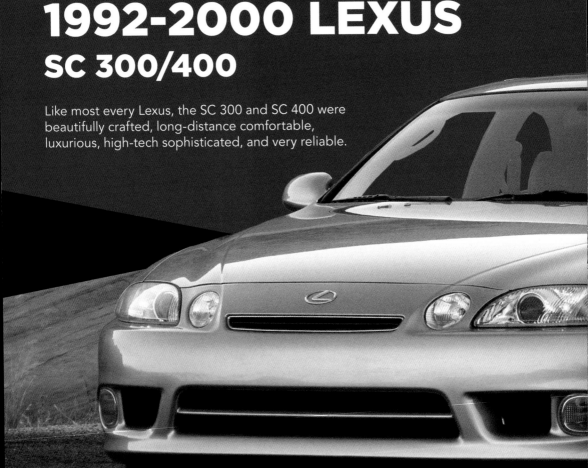

1992-2000 LEXUS
SC 300/400

Like most every Lexus, the SC 300 and SC 400 were
beautifully crafted, long-distance comfortable,
luxurious, high-tech sophisticated, and very reliable.

Wheelbase was 4.9 inches shorter at 105.9 than the Lexus LS sedan.

Lexus said the SC 400
weighed 3575 pounds
at the curb, 184 less
than the LS. The SC 300
scaled 3494 pounds.

Six-cylinder and V-8 engines were planned from the start. The former was a 3.0-liter inline unit similar to the contemporary Toyota Supra six, while the latter was the new 4.0-liter LS sedan motor. Both were all-aluminum designs with twin overhead camshafts operating four valves per cylinder. The six delivered 225 bhp and 210 pound-feet of torque. The V-8 claimed 250 bhp and 260 pound-feet.

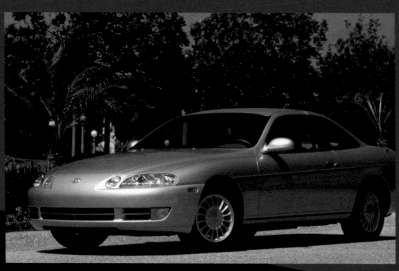

In 1998, SC borrowed powertrains from Lexus' redesigned GS midrange sedans. That brought SC 400 outputs to 300 bhp and 310 pound-feet of torque. A five-speed automatic was also new for the V-8 model. The SC 300 stayed with a four-speed automatic, but lost its manual gearbox to lack of interest, though another 10 pound-feet of torque—220 in all—was some consolation.

1999 SHELBY
SERIES 1

The Shelby Series 1 was Carroll Shelby's only clean-sheet design. (His other cars were modifications of designs from other manufacturers.) It was also the last hurrah from the storied Texas chicken farmer who won Le Mans, fathered the almighty Cobra, engineered Le Mans triumphs for Ford, and inspired the Viper.

The Series 1 was ten inches shorter than a Viper on an identical 96.2-inch wheelbase and was three inches wider than a C5 Corvette at 76.2. Base curb weight was 2650 pounds, fifty pounds under that of a base BMW Z3.

The carbon-fiber body weighed only 130 pounds, yet was stronger than a steel shell would have been.

The Series 1 was claimed to do 0–60 mph in 4.4 seconds, 0–100 in 11 flat, and a 12.8-second quarter-mile at 109.9 mph.

1996 MCLAREN

F1

Drawing on multiple Formula 1, CanAm, and Indianapolis victories, England's McLaren organization set about creating its first road car in 1989. Revealed to the public in 1992, the McLaren F1 was on the road by 1994 and in the winner's circle at LeMans in 1995. With its no-holds-barred engineering, the F1 redefined the term "supercar." When production ceased in 1997, only 100 cars, including GTR and LeMans competition versions, had been built.

K8 MCL

Scissor-type doors provided access to a leather interior with an unusual "1+2" layout: a form-fitting driver seat was centrally located, with a passenger seat slightly aft on both sides.

A BMW-designed 6.1-liter V12 was mounted amidships and packed 627 bhp. A carbon-fiber body/chassis structure made for an unprecedented power-to-weight ratio of under four lbs per horsepower.

Price and performance were equally stratospheric: $810,000, 0-60 in 3.2 seconds, 11.1-second quarter-mile times, and a 231-mph top speed.

1995 CHRYSLER
ATLANTIC

The 1995 Chrysler Atlantic was inspired by pre-World War II sport coupes, including the French Bugatti Type 57 Atlantic.

Like the Bugatti, the Chrysler was powered by a dual-overhead-cam straight-eight engine.

Chrysler combined two Dodge Neon four cylinders to create an engine layout the corporation hadn't used since the Fifties.

Unfortunately, the 1995 Chrysler Atlantic concept was not practical to manufacture because the cost of production was just plain too high. For that reason, the project was ultimately canceled.

1992 JAGUAR
XJ220

The Jaguar XJ220 drew much praise—and collected more than 1000 order deposits—when it was first shown in 1988 as a concept vehicle. That original XJ220 concept boasted a V-12 engine and projected top speed of 220 mph, but much had changed by the time the car went on sale in 1992 as a production model.

The V-12 was replaced by a 540-horsepower 3.5-liter V-6 version of the engine that powered Jaguar racecars, and all-wheel drive was replaced by rear-wheel drive.

A production XJ220 was timed at 217.1 mph—not quite 220, but fast enough to take the Guinness World Record as the fastest standard production car. The 0-60 mph time was a blistering 3.6 seconds.

In spite of the XJ220's high performance, sales were disappointing—only 271 examples were built. Many potential buyers didn't like the substitution of a V-6 engine, and an early-Nineties recession significantly hampered the supercar market.

Auction at Sotheby's Paris.

The car is shown off at a 2015 car show.

1991 VECTOR
W-8

Operating on the fringes of credibility was Vector Aeromotive Corporation, which raised $20 million in 1988 to build a Ferrari-beating mid-engined supercar but was suspiciously broke within a few years after producing fewer than 30 automobiles. Subsequent owners did produce a V12 Vector, but neither car nor company would last.

The W2 Twin Turbo pictured here is typical of the variations on a theme Vector periodically unveiled to keep press and public baited.

The W-8 purported to use military aircraft technology and had an aluminum honeycomb structure, composite body, and a twin-turbo 5.7-liter V8 of a claimed 600 bhp.

1985 BUICK
WILDCAT

The 1985 Buick Wildcat used a version of Buick's V-6 with modifications by racing specialist McLaren.

The engine was mid-mounted and drove all four wheels. Entry required pushing a solenoid, which raised the canopy.

The futuristic cockpit was forward looking, in fact, the center console gave readings on g-force, torque, oil pressure, and aided in navigation by including a compass.

1973 CHEVROLET
AEROVETTE

Chevrolet started experimenting with mid-engined Corvettes in 1969. In 1973, Chevrolet displayed a mid-engined Aerovette concept car on the auto show circuit.

The 1973 Aerovette concept was said to be very close to the automaker's planned 1980 production derivation. In the end, Chevrolet decided to continue with the conventional front-engine Corvette. A mid-engine Corvette was not to be—at least not in the Eighties.

Finally, in 2020, the dream of a mid-engined 'Vette was realized. The C8 was an unmitigated success and managed to live up to its colossal expectations.

1970 LANCIA
STRATOS HF ZERO

Italian coachbuilder Bertone took the wedge-shape, mid-engined sports car design to the limit with the 1970 Lancia Stratos HF Zero concept car.

In spite of its extreme design, the Stratos was drivable with power provided by a Lancia V-4 engine.

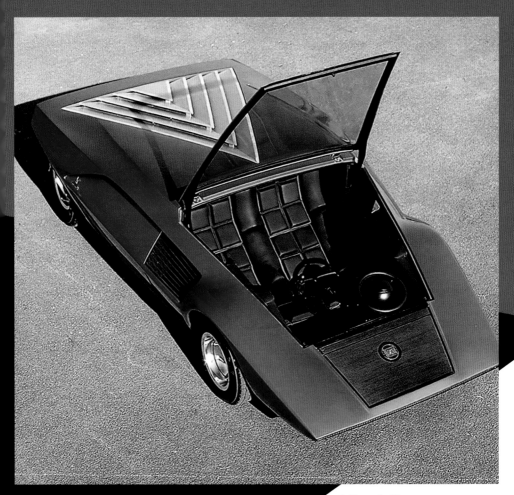

Passengers entered the Stratos by lifting the windshield. Lancia later built a more conventional production Stratos that did well in the World Rally Championship series.

1970 CHRYSLER
CORDOBA DEL ORO

Elwood Engel designed the 1961 Lincoln Continental before coming to Chrysler. Similar to the Continental, the 1970 Chrysler Cordoba Del Oro was a large car with clean lines.

A cantilevered roof allowed extremely thin windshield pillars and the rear spoiler raised automatically.

290

The Cordoba de Oro was a very wedge-shaped car. Low, long, and sleek, it still looks terrific today.

The Cordoba name would later be used five years later on Chrysler's personal luxury coupe.

1969 HOLDEN
HURRICANE

The Hurricane did not feature conventional doors; instead, a hydraulically powered canopy swung forward over the front wheels and the seats rose up and tilted forward.

293

Advanced features included on-board navigation
and a camera instead of a rear window.

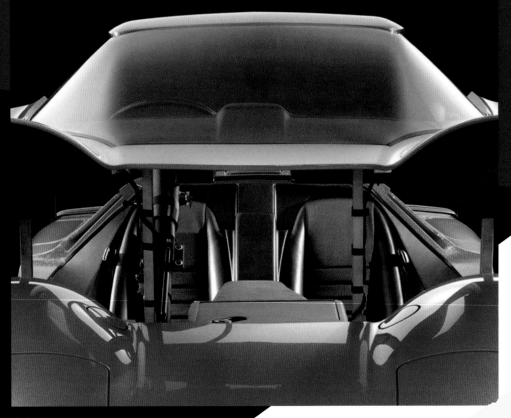

Although the Hurricane never entered production, its V-8 engine was built by Holden for many years.

1968 DODGE
CHARGER III

The 1968 Dodge Charger III concept car was far different from the production Dodge Charger. Instead of a muscle car, the Charger III seemed to be aimed at the Chevrolet Corvette market.

The exhaust exited through rectangular outlets under the rear lights, at the center of the car. The Dodge Charger III's headlights were hidden.

Like many concept cars of the Fifties and Sixties, the Charger had a canopy instead of conventional doors.

Braking was aided by three flaps in the rear that could deploy as air foils.

1964 PORSCHE
904

The 904 GTS was conceived in 1962 as Porsche's latest GT-class endurance racer but could be used on the road given a skilled, tolerant driver. A sturdy box-rail chassis with 90.6-inch wheelbase and all-independent suspension supported a smooth fiberglass body shaped by "Butzi" Porsche, grandson of the great Ferdinand and designer of the milestone 911, which also broke cover in 1963.

In 1964 alone, team entries ran 1st and 3rd in the Targa Florio, 3rd overall in the Nurburgring 1000 Kilometers, 3rd through 6th in the Tour de France, and were five of the top 12 finishers at LeMans.

A five-speed 911 transaxle teamed with an engine sitting just behind the two-seater cockpit. The package was sized for the legendary 2.0-liter Carrera flat-four from the outgoing 356 series, but some 904s got six-cylinder 911 engines, and one or two were built with racing flat-eights. Though a bit heavier than planned, the 904 was fast enough and could go the distance.

The car can be seen at the Porsche Museum in Stuttgart, Germany.

Only 100 were built over two years. A
good many survive today and are still
going strong on road and track alike.

1953 CUNNINGHAM

C-3

Briggs Cunningham devised the touring C-3 mainly to qualify his racecars as "production"– and fend off scrutiny from the IRS. Only 27 were built in 1953–55: nine convertibles and 18 coupes. All used the basic C-2R chassis, Chrysler hemi V-8, and handsome, well-appointed Vignale bodies designed by Giovanni Michelotti.

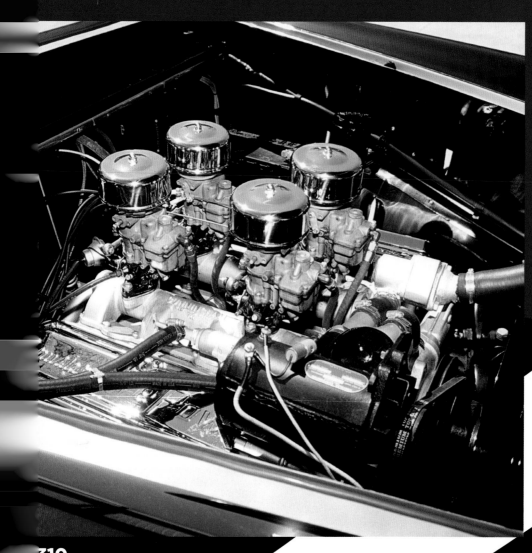

Despite eye-popping $9000-$10,000 prices, demand for the C-3 exceeded Briggs' ability to supply.

MASERATI
A6GCS

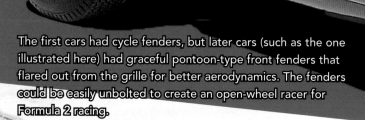

The first cars had cycle fenders, but later cars (such as the one illustrated here) had graceful pontoon-type front fenders that flared out from the grille for better aerodynamics. The fenders could be easily unbolted to create an open-wheel racer for Formula 2 racing.

The Maserati A6GCS was an ambitious combination racecar and sports car. The 2.0-liter overhead-cam inline six sat in a tubular-steel frame and initially delivered 120 bhp. But that was no match for Ferrari in European formula racing, so Maserati added a twincam head. An improved "Series II" A6GCS with modern slab-sided styling appeared in 1953.

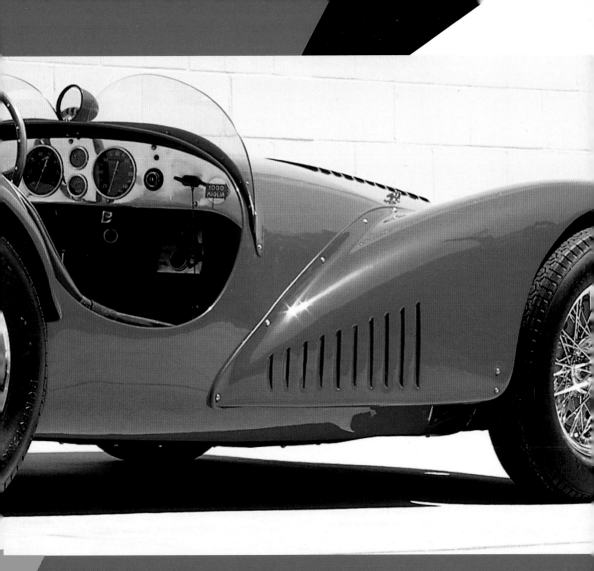

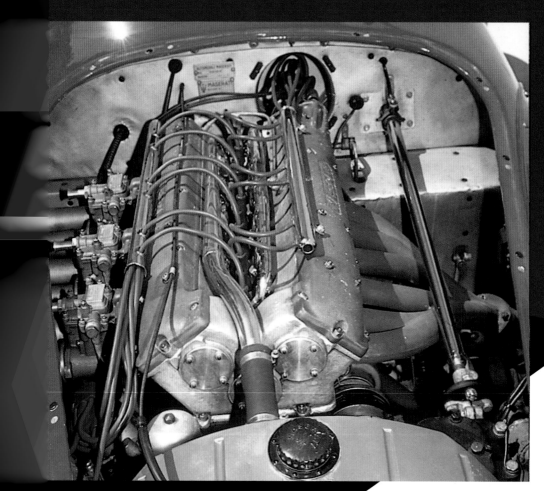

With triple Weber carburetors and two spark plugs per cylinder, the engine now developed 160 horsepower.

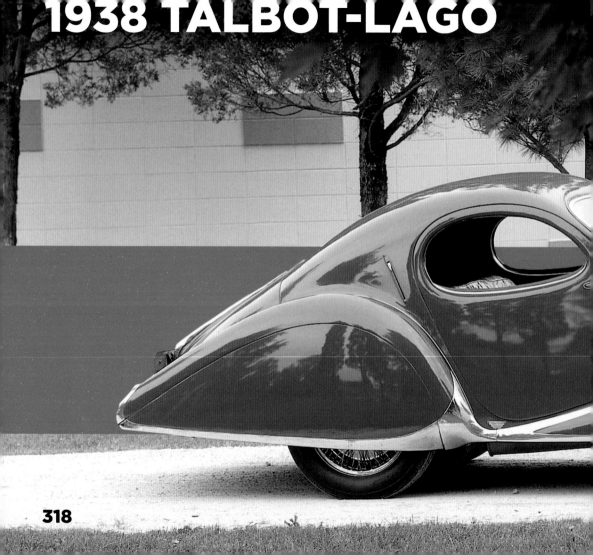